Multiple Choice Questions in Physiology

with answers and explanatory comments

Lynn Bindman
Brian Jewell
Laurence Smaje

The Department of Physiology,
University College London

ALPHA 91
MEDICAL YEAR CLUB

Edward Arnold

Introduction

Multiple-choice questions are being used increasingly in examinations at all levels, but they often test little more than factual recall. The main purpose of this booklet is to provide a selection of questions that also test reasoning power and ability to interpret data or perform simple calculations. Most of the questions were devised originally for examining preclinical medical students at University College London and they have been found to discriminate well among students of differing abilities as judged by other criteria.

We hope that teachers of physiology and their students will benefit from this booklet. Teachers may find the questions provide a useful source of ideas for creating their own questions, while students will gain experience of multiple-choice questions that test understanding as well as factual knowledge. Students may also find working through the booklet a useful form of revision as explanatory comments are provided for most of the answers; these are more detailed for points that students tend to find difficult.

How to use the booklet. The main topics covered are set out on the Contents page. Questions on the central nervous system and special senses are not included because these topics are taught and examined in a separate Neurosciences course at University College. Questions on applied and clinical physiology that are more suitable for 2nd year medical students are marked with an asterisk. Every question consists of a stem and five statements, each of which must be judged '**True**' or '**False**'. We suggest you answer all parts of the question before looking at the answers and explanations given on the opposite page.

Test yourself. You may like to compare your performance with that of our medical students. Give yourself a mark of +1 for each correct judgement, −1 for each incorrect judgement, and zero for any you left out. The score for each question can therefore vary from +5 to −5 marks. In tests consisting of 20 questions, the mean mark for University College students in examinations over a period of five years has been around +50 with a standard deviation of about 10 marks. There are of course other ways of setting and marking multiple-choice questions, and a useful discussion of these with references will be found in the booklet 'An Introduction and Guide to the Use of Multiple-Choice Questions in University Examinations' published by the University of London.

Ambiguous questions. It is extremely difficult to set questions that contain no ambiguities and for several years we have invited our examination candidates to qualify their answers to questions they find ambiguous by writing on the question paper. In fact this has necessitated very few adjustments to the marks awarded, but it has allowed us to refine some of our questions and it has undoubtedly helped to reduce the tension often induced by multiple-choice examinations. We would be pleased if users of this booklet would draw our attention to ambiguities in the questions included in it.

Acknowledgements. We thank the University of London for permission to publish some of our examination questions and Deana MacCormack for her skill and patience in typing this booklet.

Contents

Abbreviations

Units

g	gram
l	litre
m	metre
s	second
mol	mole
osmol	osmole
Pa	pascal (unit of pressure = 1 newton/m^2)
mm Hg	millimetre of mercury
	1 mm Hg $\equiv$ 133·3 Pa

Prefixes

k	kilo	($\times 10^3$)
m	milli	($\times 10^{-3}$)
μ	micro	($\times 10^{-6}$)
n	nano	($\times 10^{-9}$)

1 The following statements give reasons for including certain ingredients in physiological bathing solutions:
   (a) The composition of a physiological bathing solution should be as close as possible to that of intracellular fluid.
   (b) Nerve axons become inexcitable if they are bathed in a Na-free Ringer's solution.
   (c) Spontaneous activity will occur in nerve fibres if the Ca^{++} concentration in the bathing medium is too high.
   (d) The membrane potential of a nerve axon is more sensitive to the concentration of potassium in the bathing medium than to the concentration of any other ion.
   (e) If a bathing solution containing bicarbonate is bubbled with pure oxygen instead of 5% CO_2/95% O_2 mixture, its pH will be too low.

2 The following are statements about the osmotic pressure of plasma:
   (a) The *total* osmotic pressure of plasma is similar to that of 0·9% NaCl.
   (b) The *total* osmotic pressure of plasma is similar to that of 0·9% glucose.
   (c) The *total* osmotic pressure of the plasma is largely due to the contributions of Na + Cl ions.
   (d) The *colloid* osmotic pressure of plasma is about 25 mm Hg.
   (e) The *total* osmotic pressure of plasma opposes the ultrafiltration of fluid from the capillaries.

3 Glycerol penetrates the red cell membrane rather slowly. Which of the following will happen when red cells are suspended in a solution of 1 mol/l glycerol in water?
   (a) The red cells will immediately undergo haemolysis.
   (b) The red cells will shrink, becoming permanently crenated.
   (c) The red cells will swell first and then shrink to become permanently crenated.
   (d) The red cells will shrink first and then swell and haemolyse.
   (e) No volume changes will take place.

4 The following are statements about human red blood cells:
   (a) Red cells are rigid biconcave discs.
   (b) Normally 10 to 20% of circulating red cells contain remnants of nuclear material.
   (c) Following haemolysis, red cells release haemopoietin which stimulates the production of more red cells.
   (d) Red cells contain carbonic anhydrase.
   (e) Red cells make a major contribution to the buffering capacity of the blood.

1 (a) **False** It should resemble interstitial fluid.
  (b) **True** Na is essential for the action potential.
  (c) **False** Spontaneous activity will occur if $[Ca^{++}]$ is too *low*. See the answer to 97 (c).
  (d) **True** This is because the resting potential depends mainly on the ratio of the K concentrations inside and outside the membrane.
  (e) **False** Its pH will be too high.

$$pH = pK + \log_{10} \frac{[HCO_3^-]}{[CO_2]}$$

2 (a) **True** Red blood cells will neither swell nor shrink when placed in 0·9% NaCl which is often referred to as 'physiological' saline. Both have an osmolarity of about 300 mosmol/l.
  (b) **False** The molecular weight of glucose is different from that of NaCl, so 0·9% glucose will have a different molarity. Even if the molarities of the two solutions were equal, the osmolarities would differ because NaCl dissociates in solution and exerts a higher osmotic pressure.
  (c) **True**
  (d) **True** The plasma proteins exert a very small fraction (about 1/200) of the total osmotic pressure of about 5000 mm Hg (1 mm Hg $\equiv$ 133·3 Pa).
  (e) **False** It is only the colloid osmotic pressure that opposes the loss of fluid. Small particles like ions and glucose can diffuse across the capillaries and hence can exert no transmural osmotic pressure.

3 (a) **False** There are neither chemical nor physical factors that will cause *immediate* disruption of the membrane.
  (b) **False**
  (c) **False**
  (d) **True** The osmotic gradient between the glycerol solution (1 osmol/l) and the cell contents (0·3 osmol/l) will initially cause water to leave the red blood cell. Glycerol penetrates slowly under the concentration gradient until the osmolarity of the cell contents exceeds that of the glycerol. Water will therefore re-enter the cell and eventually the swelling produced by the continual entry of glycerol and water will burst the cell membrane.
  (e) **False**

4 (a) **False** They are not rigid, and indeed undergo considerable reversible deformation as they pass through capillaries. They only appear as biconcave discs when unstressed.
  (b) **False** Normally there are 1 to 2% of reticulocytes. Much higher percentages occur when there is accelerated haemopoiesis.
  (c) **False** The main source of haemopoietin is the kidney.
  (d) **True**
  (e) **True**

*5 A blood count in a woman aged 40 gave the following picture: Hb, 110 g/l, RBC, $3\cdot0 \times 10^{12}$/l; mean cell diameter, $8\cdot2\mu$m. The following are statements about the findings:
(a) The blood picture is within normal limits.
(b) The findings are typical of iron deficiency anaemia.
(c) The findings are typical of vitamin B_{12} deficiency.
(d) This blood would carry about 150 ml oxygen/l blood.
(e) The findings are typical of someone living at high altitude.

*6 Oxygen delivery to the tissues is usually reduced in the following conditions:
(a) Sickle cell anaemia.
(b) Reduced ventilation/perfusion ratio.
(c) Severe iron deficiency anaemia.
(d) Congestive cardiac failure.
(e) Emphysema.

7 The graph shows the relationship between the saturation of haemoglobin with oxygen and the partial pressure of oxygen.

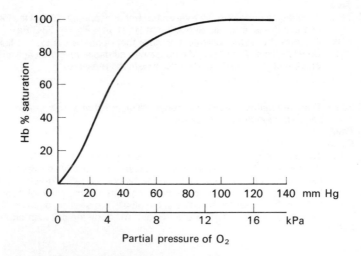

Partial pressure of O_2

(a) A rise in pH will move the curve to the left.
(b) Anaemia will depress the curve.
(c) A fall in temperature will move the curve to the left.
(d) An increase in 2, 3, diphosphoglycerate in the red cells would shift the curve to the right.
(e) Foetal haemoglobin has a similar dissociation curve.

5 (a) **False** The Hb is lower, the RBC count lower and mean cell diameter larger than normal.
  (b) **False** The woman is anaemic, but in iron deficiency anaemia red blood cells are smaller than usual.
  (c) **True** Vitamin B_{12} deficiency produces a megaloblastic anaemia which results in RBCs that are larger than normal.
  (d) **True** Each 1 g of haemoglobin can carry 1·34 ml of oxygen at NTP.
  (e) **False** In acclimatization to high altitude the red cell count and haemoglobin concentration would be higher than normal.

6 (a) **True** Sickled cells are more rigid than normal cells, and cannot therefore squeeze their way through narrow capillaries. (The abnormal Hb-S molecules form polymers when the oxygen tension is low, and the cell becomes sickle-shaped).
  (b) **True** Blood leaving the lungs is likely to be less well oxygenated when the ventilation/perfusion ratio falls.
  (c) **True**
  (d) **True** The cardiac output is reduced and the peripheral resistance vessels are likely to be constricted. Both factors reduce oxygen delivery to the tissues.
  (e) **True** As in (b) the blood is likely to be incompletely oxygenated when it leaves the lungs.

7 (a) **True**
  (b) **False** Look again at the ordinate! It is *percentage* saturation of haemoglobin, not oxygen content of the blood.
  (c) **True**
  (d) **True** This occurs for example in acclimatization and in anaemia, and assists in unloading oxygen in the tissues.
  (e) **False** It is displaced to the left, thereby facilitating oxygen transfer from mother to foetus. Foetal Hb is less affected by 2,3,DPG than adult Hb.

8 This diagram shows how the oxygen content of a sample of blood varied with the PO$_2$ of the gas with which it was in equilibrium at a PCO$_2$ of 40 mm Hg (5·3 kPa).

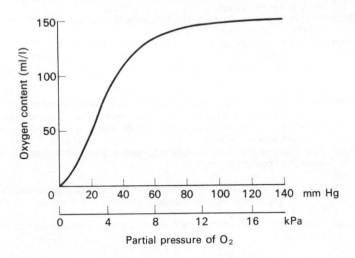

Partial pressure of O$_2$

(a) The sigmoid shape of the curve is due to the fact that red blood cells vary in their affinity for oxygen.
(b) This sample of blood must have had a haemoglobin concentration of about 150 g/l.
(c) The haemoglobin is almost fully saturated at a PO$_2$ of 100 mm Hg (13·3 kPa).
(d) If the PO$_2$ had been raised to higher values, the curve would have become horizontal when the PO$_2$ exceeded 150 mm Hg (20 kPa).
(e) If the PCO$_2$ of the sample is increased, the oxygen content would be greater at a given PO$_2$.

9 The following are statements about the viscosity of blood:
(a) Viscosity of the blood increases as the haematocrit increases.
(b) The viscosity of blood flowing through small tubes (e.g. arterioles) is lower than in large tubes (e.g. the aorta).
(c) The viscosity is less at 25°C than at 37°C.
(d) Viscosity of the blood usually increases in acclimatised mountaineers.
(e) It is decreased in people with iron-deficiency.

10 Blood clotting is delayed or prevented *in vitro* when:
(a) Blood is placed in polythene tubes (compared with glass tubes).
(b) The temperature of the blood is increased from room temperature to 37°C.
(c) Sodium citrate is added to the blood.
(d) Dicoumarol is added to the blood.
(e) When heparin is added to the blood.

8 (a) **False** It is due to the binding properties of the haemoglobin molecule for oxygen.
  (b) **False** 1 g haemoglobin can combine with 1·34 ml of oxygen at NTP.
  (c) **True**
  (d) **False** The curve (i.e. the oxygen content) would continue to rise slightly as more oxygen is dissolved in physical solution.
  (e) **False** It's the other way around (Bohr effect).

9 (a) **True** Remember that the haematocrit is the fraction of red cells in the blood sample.
  (b) **True** A surprising fact with a complicated explanation.
  (c) **False** Viscosity always increases with a fall in temperature.
  (d) **True** In acclimatization the haematocrit rises.
  (e) **True** Iron-deficiency leads to anaemia and hence to a lower haematocrit.

10 (a) **True** The non-wettable surface delays surface activation of the initial sequences of blood clotting.
  (b) **False** The optimum temperature for the enzymes involved is around 37°C.
  (c) **True** It binds the calcium ions that are essential for clotting.
  (d) **False** Dicoumarol is a competitive antagonist of vitamin K which is required for prothrombin synthesis. Dicoumarol added to the blood *in vitro* therefore has no effect on blood clotting as prothrombin is already present. *In vivo* it needs to be taken for a few days before any effects are seen.
  (e) **True** Heparin is an effective anticoagulant both *in vitro* and *in vivo* and probably acts as an anti-thrombin.

11 It is often difficult to find suitable blood for transfusion to patients who have had:
    (a) A previous transfusion of Rh+ blood.
    (b) Horse serum injections (e.g. anti-tetanus serum).
    (c) Many previous transfusions.
    (d) Syphilis or jaundice.
    (e) No previous transfusion.

12 If Rh+ blood is transfused into an Rh− woman who has not previously been transfused, then:
    (a) Anti-Rh antibodies will be produced by the woman.
    (b) The bloods are incompatible so red cell agglutination and death may follow.
    (c) In a subsequent pregnancy the foetus could be threatened by haemolytic disease.
    (d) There is no immediate or long term effect as 70% of the Rh+ population are heterozygous.
    (e) Provided anti-D antibody is given before the next pregnancy no harm will be done.

13 A man of blood group A has 2 children. Plasma from the blood of one of them agglutinates his red cells while that from the other does not.
    (a) Father *must* be heterozygous group A.
    (b) Children *must* have had different mothers.
    (c) 'Agglutinating' child *could* be group O.
    (d) Mother of 'agglutinating' child *must* be group O.
    (e) 'Non-agglutinating' child *could* be group AB.

14 The following are statements about the autonomic nerve supply to the heart:
    (a) In the normal animal there is a background level of sympathetic 'tone' to the heart.
    (b) Increased sympathetic activity decreases the rate of firing of the pacemaker cells.
    (c) Sympathetic stimulation shifts the curve relating stroke work (ordinate) to diastolic volume of the heart (abscissa) to the right.
    (d) Sympathetic stimulation increases the rate of coronary blood flow.
    (e) Stimulation of the vagus nerve slows the heart rate.

11 (a) **False** If the recipient were Rh +ve, a Rh +ve transfusion would not provoke antibody formation. If the recipient were Rh −ve, a previous Rh +ve transfusion would have provoked antibody production but Rh −ve blood is readily available, so there is no problem.

(b) **False** This could make subsequent injections of horse serum dangerous but has no effect on blood group antigen/antibody reactions.

(c) **True** Each transfusion increases the possibility of antibody formation because of subgroup incompatibility.

(d) **False** People who have had syphilis or jaundice should not be used as *donors*.

(e) **False** But in every case recipient and donor blood should be directly cross-matched even if apparently compatible.

12 (a) **True** These will be mostly anti-D.

(b) **False** There are no naturally occurring anti-Rh antibodies.

(c) **True** After the transfusion anti-Rh+ antibodies would be formed; they may cross the placental barrier and if the foetus were Rh +ve react with foetal red blood cells.

(d) **False** Irrelevant.

(e) **False** Anti-D antibody cannot be used in this situation. It can be used immediately after childbirth to prevent a Rh −ve mother from forming anti-D antibodies to cells from a Rh +ve foetus that could have passed into her circulation at parturition. The donated antibodies are destroyed within a few weeks and so will not threaten future foetuses.

13 (a) **True** The A gene is dominant, so if he were *homozygous* group A, his children would be group A or AB, and neither plasma would agglutinate his cells.

(b) **False** The mother could for example be group AB. The non-agglutinating child could be group AA and the agglutinating child, BO.

(c) **True** Or the child could be group B with B gene inherited from the mother and O from the father.

(d) **False** See answer to b.

(e) **True**

14 (a) **True**

(b) **False** It increases heart rate.

(c) **False** The curve is shifted upwards and to the left (greater stroke work at given fibre length).

(d) **True**

(e) **True**

15 An infusion of noradrenaline is given to a human subject at a sufficient rate to produce a rise in systolic BP of 20 mmHg (2·7kPa). The consequences are likely to be:
   (a) An increase in diastolic blood pressure.
   (b) Decreased firing in the baroreceptor nerves.
   (c) A reflex bradycardia (slowing of the heart).
   (d) A generalized decrease in sympathetic nerve discharge.
   (e) A decrease in [FFA] (free fatty acid concentration).

16 Cardiac output is decreased:
   (a) During stimulation of sympathetic nerves to the heart.
   (b) As a consequence of decreased pressure in the carotid sinus.
   (c) By increasing the end-diastolic volume of the heart.
   (d) On cutting the vagal nerves to the heart.
   (e) On standing up.

17 The consequences of arteriolar vasoconstriction in an organ are likely to be:
   (a) A reduction in blood flow through the organ.
   (b) An increase in capillary pressure in the vascular bed.
   (c) A decrease in the arterio-venous oxygen difference (i.e. difference in oxygen concentration in blood entering and leaving the organ).
   (d) An increase in the partial pressure of CO_2 in blood leaving the organ.
   (e) A decrease in the rate of lymph flow from the organ.

18 The following are statements about the exchange of substances between blood and the interstitial fluid:
   (a) O_2 and CO_2 pass easily across capillary walls.
   (b) The exchange of non lipid-soluble substances across capillary walls requires the presence of water-filled pores in the endothelial lining.
   (c) Movement of a substance from blood to interstitial fluid will occur only if there is a favourable concentration gradient.
   (d) Exchange diffusion provides an enormously greater turnover of water between the blood and the interstitial space than filtration and reabsorption along the lines envisaged by Starling.
   (e) Exchange of substances between blood and interstitial fluid also occurs across the walls of venules.

15 (a) **True**
   (b) **False** An *increased* firing results from the rise in blood pressure.
   (c) **True** A consequence of the increased baroreceptor discharge; vagal slowing of the heart overrides any direct cardioaccelerator action of noradrenaline.
   (d) **True** This is another reflex consequence of (b).
   (e) **False** A *rise* in [FFA] is brought by activation of β receptors. Noradrenaline acts mainly on α receptors.

16 (a) **False** Heart rate and stroke volume rise, hence cardiac output increases.
   (b) **False** There is a reflex increase in cardiac output (and peripheral resistance).
   (c) **False** It is increased by the Starling relationship except in the failing heart.
   (d) **False** The heart rate increases because of the abolition of vagal tone, hence cardiac output increases.
   (e) **True** Cardiac output drops when you stand up due to pooling of blood and remains lower in spite of compensatory reflex adjustments.

17 (a) **True** Flow is given by P/R. The vascular resistance of the organ, R, is increased by arteriolar constriction, but P, the driving pressure, will not be increased much by vasoconstriction in a single organ. Flow will therefore decrease.
   (b) **False**
   (c) **False** If the blood flow falls, *more* oxygen will be extracted during the passage of blood through the capillary.
   (d) **True**
   (e) **True** There will be diminished filtration if capillary transmural pressure is reduced.

18 (a) **True** These gases are lipid soluble and can therefore easily penetrate cell membranes.
   (b) **True** However such pores do not have to have a permanent existence.
   (c) **False** Diffusion occurs in both directions but *net* movement by diffusion can only occur *down* a concentration gradient. Note also that movement can take place *against* a concentration gradient by ultrafiltration or by active transport.
   (d) **True**
   (e) **True** Total surface area of venules approaches that of capillaries and filtration coefficients can be higher.

19 The following are important variables in circulatory physiology: cardiac output, CO; total peripheral resistance, TPR; mean arterial blood pressure, BP; stroke volume, SV; heart rate, HR. Are the following relationships true or not?
   (a) BP = CO × TPR
   (b) CO = BP/TPR
   (c) CO = SV × HR
   (d) HR = BP/(SV × TPR)
   (e) TPR = BP × SV × HR

20 The diagram shows a tube in which there is a narrow region, B, where the diameter is half that at A and C:

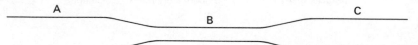

   (a) The resistance to flow (per unit length of tube) will be eight times greater at B than at A.
   (b) The velocity of flow will be four times greater at B than at A.
   (c) The lateral (side-wall) pressure at B will be greater that at A.
   (d) The pressure drop along the narrow section will be inversely proportional to its length.
   (e) The tangential stress in the wall of the tube will be less at B than at A.

21 The following are statements about veins:
   (a) Kinetic energy accounts for a much higher proportion of the total energy of blood flowing in the inferior vena cava than it does in the aorta.
   (b) Sub-atmospheric pressures are never found in blood vessels outside the thoracic cavity.
   (c) If the transmural pressure at a point A in a blood vessel is greater than that at point B, then the direction of blood flow in the vessel must be from A to B.
   (d) The pressure in the veins of the foot will be lower when a person is walking than it will be when he is standing still.
   (e) About two thirds of all the blood in the body is found in the systemic veins.

19 (a) **True**   This can be considered as a haemodynamic equivalent of Ohm's law.
   (b) **True**
   (c) **True**
   (d) **True**
   (e) **False**

20 (a) **False**   Resistance is proportional to $1/radius^4$, so if the radius is halved resistance will increase by a factor of 16.
   (b) **True**   Velocity $\times$ cross-sectional area must be the same at A and B, and the area decreases by a factor of 4.
   (c) **False**   If velocity (V) is greater at B than at A, the kinetic energy of the moving fluid ($\frac{1}{2}mV^2$) will also be greater. As the total energy of fluid (kinetic + pressure energy) cannot be greater at B than at A, pressure energy at B (and therefore side wall pressure) must be less than at A.
   (d) **False**   The pressure difference will be proportional to length (Poiseuille's Law).
   (e) **True**   The Laplace relation predicts this.

21 (a) **True**   The flow (ml/min) in both vessels is about the same, so is their cross-sectional area; therefore mean velocity and kinetic energy will be similar. However pressure energy is nearly 100 times greater in the aorta than in the inferior vena cava.
   (b) **False**   They occur in cranial sinuses, where rigid walls prevent vessels collapsing due to sub-atmospheric pressure.
   (c) **False**   *Transmural* pressure gradients have nothing to do with blood flow *along* tubes.
   (d) **True**   Due to muscle pumping and the presence of valves.
   (e) **True**

22 A and B represent two vascular beds (e.g. skin and muscle of upper limb) which are supplied with blood from the same artery (e.g. brachial). F denotes blood flow and R resistance to blood flow. Suppose the pressure drop, ΔP, across the vascular beds remains constant:

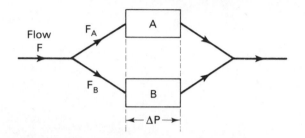

(a) The flow through A is given by: $F_A = \Delta P/R_A$
(b) The total flow is given by: $F = F_A + F_B = \Delta P/(R_A + R_B)$
(c) An increase in R_B will lead to a reduction in F_B.
(d) An increase in R_B will lead to an increase in F_A.
(e) The combined resistance, R, of the two vascular beds is given by: $R = 1/R_A + 1/R_B$.

23 Diagrams a and b are records obtained by venous occlusion plethysmography on the human forearm taken before and during (or just after) an experimental intervention:

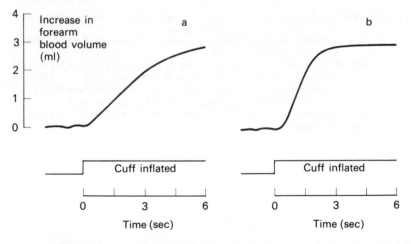

(a) Recording b could have been taken after cooling the forearm.
(b) Recording b could have been taken during fainting.
(c) Recording b could have been taken immediately after exercising the forearm.
(d) Recording b could have been taken after administration of a β-adrenergic antagonist.
(e) Recording b could have been taken by raising the occlusion cuff pressure to 60 mm Hg instead of the 40 mm Hg used in the control record a (to 8 kPa from 5·3kPa).

22 (a) **True** The plumbing version of Ohm's Law again.
   (b) **False** $F = \dfrac{\Delta P}{R_A} + \dfrac{\Delta P}{R_B} \neq \dfrac{\Delta P}{R_A + R_B}$
   (c) **True** Follows from relationship described in (a).
   (d) **False** If ΔP remains constant, F_A will not be affected by R_B. However the total flow, F, will be reduced.
   (e) **False** True statements are:

$$\frac{1}{R} = \frac{1}{R_A} + \frac{1}{R_B} \qquad R = \frac{R_A R_B}{R_A + R_B}$$

23 (a) **False** The increased slope in b results from an increased forearm blood flow (cooling would decrease flow).
   (b) **True** Fainting due to a vaso-vagal attack is accompanied by 'active' vasodilation in the vascular beds of skeletal muscle.
   (c) **True**
   (d) **False** An increased blood flow could result from abolition of vasoconstrictor tone, but that would require an α-adrenergic antagonist.
   (e) **False** Changing the cuff pressure alters the final level reached on the records, but not the initial slope.

24 The left ventricle has a thicker wall than the right because:
   (a) It has to eject blood against a higher pressure.
   (b) It has to eject a greater stroke volume.
   (c) It has to do more stroke work.
   (d) It has to eject blood through a narrower orifice.
   (e) It has to eject blood at a much higher velocity.

25 This diagram shows how the pressure and volume of a ventricle change with
   respect to one another during the cardiac cycle.

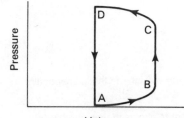

Volume

   (a) The curve between A and B is part of the passive pressure-volume relation of
       the ventricle.
   (b) Between B and C the ventricle fills with blood.
   (c) Ejection of blood from the ventricle occurs between C and D.
   (d) The area of the loop gives a good indication of the work done by the ventricle
       during one cardiac cycle.
   (e) The cardiac output can be obtained by multiplying the area of the loop by the
       heart rate.

26 The graphs show the relation between the stroke work of the left ventricle and the
   pressure in the ventricle at the end of diastole.

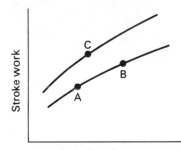

End-diastolic pressure

   (a) The end-diastolic pressure gives an indication of the size of the ventricle at the
       beginning of systole.
   (b) A change in stroke work from A to B could be explained by Starling's Law of the
       Heart.
   (c) The two graphs show different contractile states of the ventricle.
   (d) A change in stroke work from B to C could be produced by stimulating the
       sympathetic nerve supply to the heart.
   (e) A change in stroke work from A to C could be produced by a change in posture
       from lying to standing.

24 (a) **True** Remember that pressures in the pulmonary artery are about 1/6 of those in the aorta. A higher pressure requires greater (tangential) wall stress and more muscle fibres in parallel are needed to produce this.

(b) **False** Stroke volumes of R and L ventricles must be identical in the long run.

(c) **True** Main component of stroke work = stroke volume × mean pressure in outlet vessel during ejection

(d) **False**

(e) **False** Consider answers to b and d: If the same stroke volume is ejected through the same sized orifice, the mean velocity will be the same for both ventricles.

25 (a) **True** Pressure rises as relaxed ventricle fills with blood.

(b) **False** Ventricle cannot be filling if no volume change occurs: this is the phase of isovolumic contraction.

(c) **True** Note that the pressure increases during ejection.

(d) **True** Area gives pressure work done during cardiac cycle (see 24c). Note that the ventricle also imparts kinetic energy to blood ejected.

(e) **False** Area × heart rate = power output (i.e. work output per unit time).

26 (a) **True**

(b) **True**

(c) **True**

(d) **True**

(e) **False** While there is an increase in sympathetic tone on standing, there is a fall in end-diastolic volume (and therefore in end-diastolic pressure).

27 These diagrams show the effect of certain procedures on the blood flow to muscle.

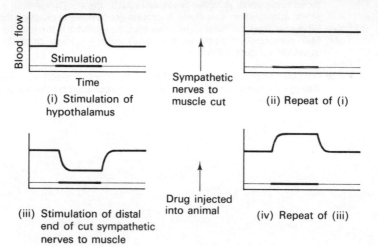

(i) Stimulation of hypothalamus

Sympathetic nerves to muscle cut

(ii) Repeat of (i)

(iii) Stimulation of distal end of cut sympathetic nerves to muscle

Drug injected into animal

(iv) Repeat of (iii)

(a) There was vasoconstrictor tone in the muscle before the nerves were cut.
(b) The sympathetic nerve supply to the muscle contained both vasoconstrictor and vasodilator fibres.
(c) Activation of sympathetic vasodilator fibres is possible by stimulation of the hypothalamus.
(d) The drug injected between (iii) and (iv) is likely to have been a β-adrenergic antagonist.
(e) The change in blood flow produced by stimulation of the hypothalamus could have been due to the systemic release of a vasoactive substance.

28 This is a schematic diagram of the relation between blood flow through a vascular bed and the pressure drop across it (ΔP).

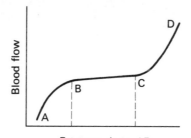

Pressure drop, ΔP

(a) This graph shows that the flow is proportional to the pressure drop across the vascular bed.
(b) This pressure-flow relation shows that autoregulation occurs in the vascular bed.
(c) This pressure-flow relation is typical of that seen in the pulmonary vascular bed.
(d) Within the region B-C, the resistance to flow increases as ΔP increases.
(e) A pressure-flow relation of the form shown is only obtainable if the autonomic nerve supply to the vascular bed is intact.

27 (a) **True** Compare the basal flows in (ii) and (i).
  (b) **True** Absence of effect of stimulation in (ii) shows that the sympathetic nerves included vasodilator fibres: the occurrence of effect (iii) shows that sympathetic nerves included vasoconstrictor fibres.
  (c) **True** See (i) and (ii).
  (d) **False** The drug has prevented vasoconstriction and is therefore likely to have been an α-adrenergic antagonist.
  (e) **False** Procedure (ii) showed that the cause of the vasodilatation in (i) was neural not humoral.

28 (a) **False** A proportional relationship must be linear: in fact the curve is S-shaped.
  (b) **True** i.e. blood flow varies very little with pressure over the range B-C.
  (c) **False** In the pulmonary vascular bed blood flow increases with very little change in pressure over the physiological range. In the diagram opposite it is *flow* that remains fairly constant.
  (d) **True** (Can you suggest a possible mechanism?)
  (e) **False** It can be found in the isolated artificially-perfused kidney.

29 This is a schematic diagram of a typical electrocardiogram from lead II for a single heart beat.

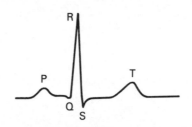

(a) The P wave is produced by depolarisation of the atria.
(b) The QRS complex is produced by depolarisation of the ventricles.
(c) The Q-T interval gives a rough indication of the duration of ventricular systole.
(d) The first heart sound occurs at about the same time as the P wave.
(e) The second heart sound occurs at about the same time as the QRS complex.

*30 Study the accompanying electrocardiograms and consider whether the following statements are true.

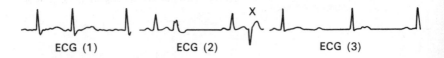

ECG (1)   ECG (2)   ECG (3)

(a) ECG (1) shows sinus arrhythmia.
(b) ECG (2) shows atrial fibrillation.
(c) X in ECG (2) is a ventricular extrasystole.
(d) ECG (3) shows complete heart block.
(e) The patient from whom ECG (3) was taken would be unable to increase his heart rate substantially on exercise.

31 Which of the following are necessary to calculate cardiac output using the direct Fick method based on oxygen measurements?
(a) O_2 consumption per minute.
(b) Arterial O_2 concentration.
(c) Ventilation/perfusion ratio.
(d) Mixed venous O_2 concentration.
(e) Diffusing capacity for O_2.

29 (a) **True**
   (b) **True**
   (c) **True**
   (d) **False** The first sound occurs as the ventricles contract and is mainly due to closure of the A-V valves.
   (e) **False** The *first* heart sound occurs at this time. The 2nd sound is due to closure of the aortic and pulmonary valves, and occurs at about the same time as the T wave.

30 (a) **False** It shows atrial fibrillation. (Sinus arrhythmia is a changing heart rate during the respiratory cycle due to autonomic influences on the sino-atrial node.)
   (b) **False** It shows a mixture of ventricular extrasystoles (beginning and end) and normal complexes (middle).
   (c) **True**
   (d) **True** P waves are seen to be separate from QRS complexes.
   (e) **True** This person is suffering from complete heart block, and the ventricular pacemakers are only slightly affected by catecholamines.

31 (a) **True**
$$\text{Cardiac output (l/min)} = \frac{\text{Oxygen consumption (ml/min)}}{\text{A-V difference (ml } O_2\text{/l blood)}}$$
   (b) **True**
   (c) **False**
   (d) **True** This needs to be genuinely mixed, i.e. from right ventricle or pulmonary artery. Peripheral venous blood will not do.
   (e) **False**

32 During prolonged exercise the following cardiovascular adjustments take place.
  (a) Muscle blood flow increases partly due to increased parasympathetic vasodilator nerve discharge.
  (b) Muscle blood flow increases partly as a consequence of the local release of vasodilator metabolities.
  (c) Skin blood flow increases.
  (d) Pulmonary vascular resistance decreases.
  (e) The increase in cardiac output observed on exercise is largely brought about by the Starling mechanism.

33 If you attempt to expire forcibly against a closed glottis, the intrathoracic pressure may rise as high as +100 mm Hg (13·3 kPa). Which of the following will be sustained effects of this manoeuvre?
  (a) A rise in intratracheal pressure.
  (b) A rise in right ventricular output.
  (c) A reduction in left ventricular output.
  (d) A fall in systemic arterial pressure.
  (e) A fall in heart rate.

*34 The following are statements about the clinical assessment of the cardiovascular system:
  (a) Atrial fibrillation results in an irregular radial pulse.
  (b) A pulse rate of 40/min would suggest complete heart block.
  (c) Sphygmomanometry by the palpatory method is useful for estimating the patient's diastolic blood pressure.
  (d) An essential preliminary to sphygmomanometry by the auscultatory method is to place the bell of the stethoscope so that the arterial pulse can be clearly heard before the cuff is inflated.
  (e) The jugular vein normally fills to about 20 mm above the sternal notch when the patient is in the sitting position.

*35 Cyanosis is a dusky bluish colouration of mucous membranes and/or skin which occurs when the blood there contains 50 g/l or more of deoxygenated haemoglobin. This *could* occur as a result of:
  (a) Reduction in the pulmonary diffusing capacity for oxygen.
  (b) Shunting of blood from the left side of the heart to the right (e.g. through a patent formamen ovale).
  (c) Too high a ventilation-perfusion ratio in the lungs.
  (d) Peripheral vasoconstriction.
  (e) Increased red cell count (polycythaemia).

*36 Acute left ventricular failure is accompanied by the following:
  (a) Increase in pulmonary blood volume.
  (b) Normal jugular venous pressure.
  (c) Generalised vasodilatation.
  (d) Ankle oedema.
  (e) Raised systemic arterial pressure.

32 (a) **False** Definitely not: there is no general peripheral distribution of parasympathetic vasodilator nerves.
   (b) **True**
   (c) **True** This helps to reduce body temperature which is raised by the heat produced by active muscle.
   (d) **True** As cardiac output increases (see answer to 28c).
   (e) **False** End-diastolic volume may increase during exercise in the erect posture, but the increase in cardiac output is the due mainly to the effect of the sympathetic nervous system on the heart.

33 (a) **True**
   (b) **False** Right ventricular output falls because high intrathoracic pressure greatly reduces venous return.
   (c) **True** This results from the fall in right ventricular output.
   (d) **True** This results from c.
   (e) **False** Fall of systemic blood pressure leads to compensatory increase in heart rate.

34 (a) **True**
   (b) **True** Note however that very fit young people may have resting heart rates as low as this.
   (c) **False** Only the systolic pressure can be estimated by this method.
   (d) **False** No sound can be heard until the cuff is inflated sufficiently to cause partial occlusion of the vessel.
   (e) **False** That would indicate a raised venous pressure.

35 (a) **True**
   (b) **False** The shunt is *left* to *right*. (This eventually leads to heart failure but initially the blood is fully oxygenated.)
   (c) **False**
   (d) **True** Slow flow leads to removal of more oxygen from blood.
   (e) **True** Slow flow again, but this time due to increased blood viscosity.

36 (a) **True** The rise in pulmonary vascular pressure may lead to pulmonary oedema.
   (b) **True** Jugular venous pressure is raised in *right* ventricular failure.
   (c) **False**
   (d) **False** This follows chronic right ventricular failure.
   (e) **False** Left ventricular output is decreased.

*37 The following might be expected to occur in a patient suffering from right ventricular failure of the heart:
   (a) Raised central venous pressure.
   (b) Enlargement of the liver.
   (c) Pulmonary oedema.
   (d) Reduced systemic arterial blood pressure.
   (e) Decreased aldosterone plasma concentration.

*38 A blood pressure of 180/120 mm Hg (24/16kPa) was found in a patient aged 35. The following are statements about this:
   (a) Since the blood pressure is raised there must be a corresponding rise in cardiac output.
   (b) Renal artery stenosis could produce this finding.
   (c) Phaeochromocytoma could produce this finding.
   (d) The figures quoted happen to be at the upper end of the range of blood pressures found in the healthy population of that age group.
   (e) The left ventricular stroke work is increased because of the high blood pressure.

*39 Peripheral circulatory failure due to haemorrhage is accompanied by:
   (a) Raised haematocrit.
   (b) Rapid heart rate.
   (c) Cold pale clammy skin.
   (d) Reduction in central venous pressure.
   (e) Increased reticulocyte count.

40 Which of the following are consequences of breathing 5% CO_2 in air?
   (a) There is a rise in the PCO_2 of mixed venous blood.
   (b) The pH of arterial blood increases.
   (c) There is an increase in alveolar ventilation.
   (d) The oxygen dissociation curve of haemoglobin is shifted to the left.
   (e) There is a decrease in cerebral blood flow.

37 (a) **True**
  (b) **True**
  (c) **False** That requires raised pressure in the pulmonary capillaries, which is a feature of *left* heart failure.
  (d) **False**
  (e) **False** The aldosterone level usually rises in chronic failure.

38 (a) **False** Raised blood pressure could be due to peripheral vasoconstriction from a variety of causes.
  (b) **True** Renin is released from the affected kidney. This leads to hypertension through the renin-angiotensin-aldosterone mechanism.
  (c) **True** This is a tumour of the adrenal medulla.
  (d) **False**
  (e) **True**

39 (a) **False** The haematocrit falls as extracellular fluid moves from the interstitium into the vascular compartment.
  (b) **True** Due to increased sympathetic activity.
  (c) **True** This is due to a combination of vasoconstriction and sweating, both of which are also consequences of sympathetic activity.
  (d) **True** Central venous pressure is useful for monitoring extracellular fluid volume.
  (e) **False** This will occur in a few days if the patient survives.

40 (a) **True**
  (b) **False** Increased PCO_2 results in an increased carbonic acid concentration in the blood and this will cause a fall of pH.
  (c) **True**
  (d) **False** The curve is shifted to the right – i.e. Hb carries less O_2 at a given PCO_2.
  (e) **False** Cerebral blood flow is increased. Arterial blood PCO_2 is the main factor controlling cerebral blood flow.

41 The following are statements about the consequences of voluntarily increasing ventilation three-fold. (Note that the isoelectric point for plasma proteins is about pH 5.)
    (a) The alveolar PO_2 trebles.
    (b) The plasma proteins become more ionised.
    (c) The plasma ionised calcium level increases.
    (d) A fall in the plasma bicarbonate concentration in arterial blood.
    (e) A brief period of apnoea immediately afterwards.

42 The diagram shows the relation between pulmonary ventilation and alveolar PCO_2 at two different values of alveolar PO_2. Are the following statements correct deductions from the data in the graph?

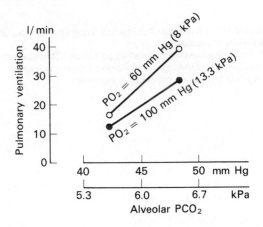

    (a) There is a direct relationship between pulmonary ventilation and alveolar PCO_2.
    (b) There is an inverse relationship between pulmonary ventilation and alveolar PO_2.
    (c) The increment in pulmonary ventilation produced by a given rise in alveolar PCO_2 is less when the alveolar PO_2 is raised from 60 to 100 mm Hg (from 8 to 13 kPa).
    (d) The straight line joining the data points for PO_2 = 60 mm Hg (open circles) is fitted by the equation:
$$\text{Pulmonary ventilation} = 4 \times \text{alveolar } PCO_2$$
$$\text{(l/min)} \qquad\qquad \text{(mm Hg)}$$
    (e) When pulmonary ventilation increases, the alveolar PCO_2 rises.

43 The following measurements were made on a patient:
        Tidal volume 500 ml, Dead space 100 ml,
        Rate of breathing 15/min, Cardiac output 7 litres/min.
    Are these statements correct?
    (a) The pulmonary ventilation is 7·5 litres.
    (b) The subject is likely to be under 8 yrs of age.
    (c) The alveolar ventilation is 6 litres/min.
    (d) The ventilation/perfusion ratio is the pulmonary ventilation divided by the cardiac output.
    (e) The ventilation/perfusion ratio is 0·86.

41 (a) **False** Maximum PO_2 in the airways if air is being breathed is 21% of atmospheric pressure minus SVP_{H_2O} at body temperature – e.g. $0.21 \times (760 - 47) = 150$ mm Hg. The value will be lower still in the alveoli because of the PCO_2.

 (b) **True** Hyperventilation will raise the plasma pH (respiratory alkalosis), thus taking the plasma proteins further from their isoelectric point.

 (c) **False** Although the total plasma [Ca] is unaffected, the *proportion* in ionized form decreases because of increased binding of calcium to various substances when pH rises. This may lead to tetany.

 (d) **True** Bicarbonate concentration falls because the CO_2 concentration falls, but note that the *ratio* $\dfrac{[HCO_3^-]}{[CO_2]}$ rises.

 (e) **True** Apnoea means absence of breathing; this is a consequence of the lowered arterial PCO_2.

42 (a) **True** At a given PO_2, increase of PCO_2 is associated with an increase in pulmonary ventilation.

 (b) **True** At a given PCO_2, increase of PO_2 is associated with a decrease in pulmonary ventilation.

 (c) **True** Slope of line is reduced.

 (d) **False** The slope is given correctly, but the equation should be $PV = (4 \times PCO_2) - C$ (value of C is about 150 l/min).

 (e) **False** The graph shows pulmonary ventilation as the dependent variable and PCO_2 as the independent variable. These cannot be transposed when dealing with homeostatic mechanisms.

43 (a) **False** It should be 7·5 litres *per minute*. (Errors of this kind are often missed by intelligent people: the moral of the story is that you must always check the units.)

 (b) **False** They are typical values for a 70 kg man.

 (c) **True** $15 \times (500 - 100) = 6000$ ml/min.

 (d) **False** It should be *alveolar* ventilation divided by cardiac output.

 (e) **True** Follows from (d).

44 In the experiment illustrated below, the respiratory responses to hypercapnia and hypoxia are shown before and after denervation of the carotid and aortic bodies. Which of the following are correct statements on the basis of the experimental results?

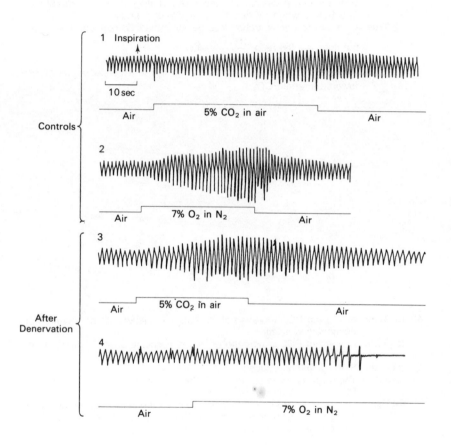

(a) Peripheral chemoreceptors (aortic and carotid bodies) are required for the respiratory response to CO_2 excess.
(b) There are chemoreceptors other than those in the aortic and carotid bodies.
(c) Peripheral chemoreceptors are required for the response to O_2 lack.
(d) The denervation results in slowing of the respiratory rate.
(e) The effects seen in (4) are sometimes observed in normal humans (i.e. with chemoreceptors intact).

44 (a) **False** There is still a response to breathing 5% CO_2 after denervation (Graph 3).

    (b) **True** Follows from (a).

    (c) **True** The response to breathing 7% O_2 after denervation is a depression of respiration (Graph 4).

    (d) **True** Probably due to cutting vagal afferents along with the chemoreceptor nerves, thus removing inhibitory drive to inspiration.

    (e) **False** Although some humans do not hyperventilate in response to a slight reduction in inspired PO_2, hyperventilation is always observed at values below 10%.

*45 The diagrams show spirometer tracings obtained from an adult subject in whom tidal volume, FEV$_1$ and vital capacity were measured before and after inhalation of isoprenaline.

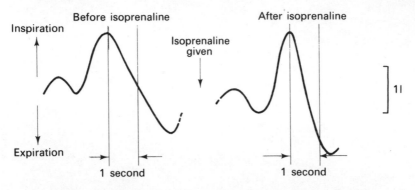

(a) Before isoprenaline, the FEV$_1$ was about 70% of vital capacity.
(b) The tracing obtained before isoprenaline was given was within normal limits.
(c) The vital capacity was increased by about 20% after inhalation of isoprenaline.
(d) Expiratory flow rate after inhalation of isoprenaline was consistent with a fall of airway resistance.
(e) These tracings are typical of those expected from an asthmatic patient.

46 These graphs are volume/pressure plots for the lungs during inspiration and expiration. Compared with the normal situation illustrated by graph A, the other graphs *could* be produced by:

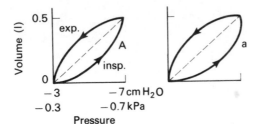
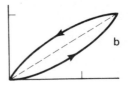
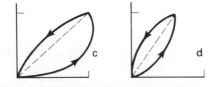
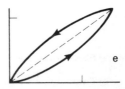

(a) An increased viscous work of breathing.
(b) A decreased lung compliance.
(c) An increase in the rate of inhalation.
(d) A diminished tidal volume.
(e) An increased elastic work of breathing.

45 (a) **False** FEV_1 was about 40% of vital capacity.
   (b) **False** FEV_1 should normally exceed 80% of vital capacity.
   (c) **True**
   (d) **True** Expiration occurred faster because isoprenaline relaxes smooth muscle of the airways.
   (e) **True**

46 (a) **True** Area of loop is increased.
   (b) **True** Bigger pressure change required for same volume change.
   (c) **True** Work of inspiration for same change of volume has increased.
   (d) **False** Tidal volume is same as in A.
   (e) **True** Area between dashed line and ordinate (= elastic work) is greater than in A.

47 The following are statements about the surface-active material (surfactant) lining
    the lung alveoli:
    (a) The surfactant increases the surface tension of the film of liquid lining the
        alveoli.
    (b) Surfactant increases lung compliance.
    (c) The surface tension of fluid containing surfactant increases as the surface area
        of the fluid decreases.
    (d) In the absence of normal surfactant there will be a greater tendency for alveoli
        to collapse.
    (e) An indirect consequence of absence of normal surfactant may be a fall in
        systemic arterial pH.

*48 Analyses of gases in alveolar air and systemic arterial blood were made on a
     patient, and were:
         Alveolar air PO_2 102 mm Hg (13·6 kPa); Systemic arterial blood PO_2
         70 mm Hg (9·3 kPa).
     Are the following statements based on these values correct?
     (a) They are typical values for a healthy patient.
     (b) They could be explained by a reduced diffusing capacity for oxygen.
     (c) The values are typical for a healthy person who lives at high altitude.
     (d) They could be explained by a shunt of deoxygenated blood into systemic
         arterial blood.
     (e) The patient may have been hypoventilating.

49 The following are statements about the ventilation and perfusion of alveoli with
    blood.
    (a) Overperfusion of normally ventilated lungs leads to a lowered PCO_2 in sys-
        temic arterial blood.
    (b) Overperfusion of normally ventilated lungs leads to a lowered PO_2 in systemic
        arterial blood.
    (c) Overventilation of some alveoli that are normally perfused can compensate
        for changes in systemic arterial PCO_2 caused by overperfusion of other alveoli.
    (d) If alveoli are underperfused and normally ventilated the blood leaving them will
        have a normal oxygen content.
    (e) If some alveoli are both underperfused and underventilated, the gas tensions
        in the pulmonary capillary blood could well have normal values.

50 The following statements concern pulmonary blood flow:
    (a) The pulse pressure in the pulmonary artery is about the same as that in the
        aorta.
    (b) If the blood pressure in pulmonary capillaries is 5 mm Hg (0·7 kPa) above the
        intra-alveolar pressure, fluid will pass into the alveolar air spaces.
    (c) The ventilation/perfusion ratio is the same in all parts of the lung in standing
        man.
    (d) During prolonged bouts of coughing the venous return to the right side of the
        heart is reduced.
    (e) Deep inspiration will increase the capacity of the pulmonary capillaries.

47 (a) **False** Surfactant decreases surface tension.
 (b) **True** It reduces the force required to overcome surface tension effects when alveoli expand.
 (c) **False** Because the surfactant remains at the water air interface, the space between surfactant molecules decreases as the surface area is reduced; this is equivalent to raising its concentration, which lowers surface tension.
 (d) **True** This is a consequence of the raised surface tension.
 (e) **True** The greater tendency of the alveoli to collapse can lead to inadequate ventilation and a rise in alveolar PCO_2 – and hence a fall in pH of the systemic arterial blood.

48 (a) **False** The alveolar value is normal but the systemic arterial PO_2 should be much closer to it.
 (b) **True**
 (c) **False** The alveolar PO_2 would be lower, and in any case this would not account for the large difference between alveolar and systemic arterial blood values.
 (d) **True**
 (e) **False** See comment for (c).

49 (a) **False** The ventilation/perfusion ratio will be lowered. The blood in the lungs will not equilibrate with alveolar air, so that systemic arterial PCO_2 (P_aCO_2) will rise. It cannot fall in these circumstances.
 (b) **True** A similar explanation to (a), leading to a fall in P_aO_2.
 (c) **True** Regional overventilation can reduce CO_2 in blood leaving the overventilated alveoli. This reduction can compensate for the increase of CO_2 in blood from overperfused alveoli. Note that overventilation will not markedly alter the oxygen content since Hb is saturated at the normal alveolar PO_2 of 100 mm Hg.
 (d) **True** The blood will have equilibrated with the alveolar air.
 (e) **True** Abnormal distribution of the inspired gas in the alveoli (due for example to compliance or resistance changes) does not necessarily lead to abnormal gas exchange. Compensatory mechanisms can match blood flow to the ventilation.

50 (a) **False** Systolic and diastolic pressures in the pulmonary artery are about 1/6 of those in the aorta and so is the pulse pressure.
 (b) **False** The colloid osmotic pressure of plasma is enough to prevent this.
 (c) **False** Blood flow is relatively less in the upper parts of the lung than the lower regions, but this is not matched by the differences in degree of ventilation.
 (d) **True** Greatly increased intrapleural pressure can collapse great veins within the thorax.
 (e) **True** This results from the increased transmural pressure caused by the subatmospheric extravascular (intrapleural) pressure.

51 The following are statements about thirst:
   (a) Drinking may be provoked by electrical stimulation of certain regions of the hypothalamus.
   (b) Loss of blood volume (e.g. haemorrhage) is not accompanied by thirst if there is no change in plasma osmolarity.
   (c) Thirst is produced by a rise in plasma osmolarity even if the blood volume is normal.
   (d) If a dehydrated person is allowed access to water, thirst and drinking continue unabated until the plasma osmolarity is restored to normal.
   (e) The sensation of thirst is related to the rate of resting salivation.

52 The following are statements about glomerular filtration:
   (a) The glomerular filtrate is produced by essentially the same mechanisms as interstitial fluid.
   (b) Glomerular filtrate has the same composition as lymph collected from the thoracic duct.
   (c) Blood in the efferent glomerular arteriole (i.e. the one carrying blood away from the glomerulus) is more viscous than blood in the afferent arteriole.
   (d) The glomerular filtration rate (GFR) is directly proportional to the systemic arterial blood pressure.
   (e) The glomerular filtration rate is the main factor determining the rate of urine production.

53 If the concentration of a substance X in the Plasma is P_X (mg/ml.) and in the urine is U_X (mg/ml), and the volume of urine produced per minute is V (ml), which of the following statements are correct?
   (a) The rate of excretion of substance X is $U_X.V$ (mg/min).
   (b) The quantity $U_X.V/P_X$ is the minimum volume of plasma from which the kidneys could have obtained the amount of X excreted per minute.
   (c) If X is filtered at the glomerulus and neither secreted nor absorbed in the renal tubules, then $U_X.V/P_X$ is the volume of plasma that passes through the kidney in one minute.
   (d) If X is inulin, then $U_X.V/P_X$ is the volume of glomerular filtrate produced per minute.
   (e) If the ratio U_X/P_X exceeds the ratio U_{inulin}/P_{inulin} then substance X must be secreted by the renal tubules.

51 (a) **True**
   (b) **False** Thirst is a prominent feature of a sudden reduction in blood volume.
   (c) **True**
   (d) **False** Thirst is initially slaked by drinking an amount of water that is insufficient to restore plasma osmolarity to normal; subsequently thirst returns, and drinking occurs once more.
   (e) **True**

52 (a) **True** Both are formed by ultrafiltration.
   (b) **False** It contains only traces of protein, whereas lymph contains appreciable amounts.
   (c) **True** The loss of 1/6 of the plasma volume as glomerular filtrate will result in increased concentrations of red cells (i.e. raised haematocrit) and plasma proteins.
   (d) **False** Glomerular capillary pressure and glomerular filtration rate are largely independent of changes in systemic arterial pressure in healthy kidneys because of autoregulation of blood flow.
   (e) **False** The rate of urine production in Man is dominated by tubular function, not by GFR.

53 (a) **True**
   (b) **True** This is a definition of the concept of renal 'clearance' of X.
   (c) **False** It is the volume of glomerular filtrate in one minute. If X were a substance such as PAH or diodrast, secreted as well as filtered so that the kidney removed all X going to it, then the clearance gives renal plasma flow.
   (d) **True**
   (e) **True** Because inulin is only filtered.

54 This diagram shows how the rates of filtration (F), secretion (S), and excretion (E) of para-amino-hippuric acid (PAH) vary with its plasma concentration (P_{PAH}).

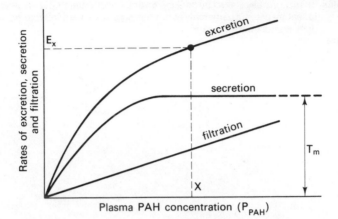

Plasma PAH concentration (P_{PAH})

(a) The rate of filtration of PAH is directly proportional to its plasma concentration.
(b) At all values of P_{PAH} the rate of excretion of PAH is given by
$$E = F + S$$
(c) T_m is the maximum tubular reabsorptive capacity for PAH.
(d) The rate of filtration of PAH is given by
$$F = k.P_{PAH}$$
where k is the slope of the filtration curve in the above diagram.
(e) The rate of excretion of PAH at plasma concentration, X, is given by
$$E_x = T_m - k.X$$

*55 In healthy adults, plasma inorganic phosphate concentration lies between 0·6 and 1·5 mmol/l and inulin clearance is typically 120 ml /min. Phosphate appears in the urine when plasma levels exceed about 0·9 mmol/l
(a) T_m phosphate is about 0·11 mmol/ min
(b) If the plasma phosphate concentration is 1 mmol/l, the amount of phosphate excreted in the urine will be about 0·12 mmol/min.
(c) Parathormone tends to reduce T_m phosphate.
(d) A low value for the ratio T_m phosphate: inulin clearance can lead to hypo-phosphataemia.
(e) A low T_m phosphate combined with a high inulin clearance is commonly found in renal failure.

54 (a) **True** Graph is a straight line through the origin.
   (b) **True**
   (c) **False** T_m (as illustrated here) is the maximum tubular *secretory* capacity. (A nasty catch question, but it emphasizes the need to read the statement word by word before making a decision.)
   (d) **True** This is an algebraic statement of (a).
   (e) **False** $E_x = T_m + k.X$.

55 (a) **True** Inulin clearance is a measure of the glomerular filtration rate so the delivery of phosphate to the renal tubules (filtered load) is:
   $$GFR \times [phosphate] = 0\cdot12 \times 0\cdot9 = 0\cdot108$$
   $$l/min \quad mmol/l \quad mmol/min$$
   Phosphate appears in the urine when the plasma concentration exceeds $0\cdot9$ mmol/l, because the filtered load then exceeds the T_m for phosphate reabsorption.
   (b) **False** Amount excreted = filtered load − amount reabsorbed (T_m)
   $$(1 \times 0\cdot12) \quad - \quad 0\cdot108$$
   $$mmol/l \ l/min \qquad mmol/min$$
   $$= 0\cdot012 \ mmol/min$$
   (c) **True** This is one of the actions of parathormone.
   (d) **True** Significant hypophosphataemia is generally seen only if T_m falls. This may be in response to a raised parathormone level ('secondary hypophosphataemia') or a direct failure of tubular reabsorption of phosphate ('primary hypophosphataemia').
   (e) **False** In chronic renal failure GFR falls and T_m falls roughly in parallel. This is widely believed to indicate that the number of active nephrons is reduced.

56 The following are statements about renal tubules:
   (a) At the tip of the loop of Henle in the renal medulla, the osmolarity of the tubular contents is several times that of the glomerular filtrate.
   (b) The walls of the ascending limb of the loop of Henle are freely permeable to water.
   (c) The fluid entering the distal convoluted tubule is hypotonic with respect to plasma.
   (d) The main force responsible for water reabsorption from the collecting tubule is the sub-atmospheric pressure in the surrounding interstitial space.
   (e) The permeability of the collecting tubule to water is under the control of aldosterone.

57 The following are statements about the acidification of urine:
   (a) H^+ is secreted into the urine by the cells lining the distal tubules.
   (b) K^+ is normally reabsorbed from the tubular fluid in exchange for H^+.
   (c) The H^+ reacts with the NaH_2PO_4 in the tubular fluid to give Na_2HPO_4.
   (d) When there is a large load of H^+ to be excreted, most of it appears in the urine in the form of ammonium salts.
   (e) H_2CO_3 is present in the urine in very high concentration compared with plasma.

*58 The diagram shows the relation between plasma bicarbonate concentration and the pH of blood at three different values of PCO_2 (60, 40 and 20 mm Hg) (8, 5·3 and 2·7 kPa). N represents the blood picture in a normal person.

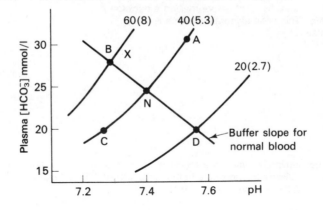

   (a) A movement along the line NB occurs in an uncompensated respiratory acidosis.

   (b) The equation of line ANC is $pH = pK' + \log_{10} \dfrac{[HCO_3^-]}{a.PCO_2}$

   where a = solubility coefficient for CO_2 and PCO_2 = 40 mm Hg (5·3 kPa).
   (c) A movement along the line NA could be produced by hyperventilation.
   (d) Point X could be reached in a partially compensated metabolic acidosis.
   (e) The slope of the line BND will be reduced if the haemoglobin concentration in the blood is decreased.

56 (a) **True** This results from counter current multiplier.
   (b) **False** They are relatively impermeable to water.
   (c) **True** Due to (b) and active removal of NaCl from lumen of ascending limb of loop of Henle.
   (d) **False** The force is osmotic, due to high osmolarity of the interstitial fluid, compared with that of the tubular contents.
   (e) **False** The permeability is controlled by antidiuretic hormone (ADH).

57 (a) **True** It is also secreted in the proximal tubules.
   (b) **False** Na^+ is reabsorbed in exchange for H^+ or K^+.
   (c) **False** Other way around.
   (d) **True** The urine smells of ammonia under these conditions.
   (e) **False** The formation of H_2CO_3 occurs by combination of secreted H^+ with HCO_3^- in the tubular fluid, but it is followed by breakdown to $CO_2 + H_2O$, and the CO_2 diffuses back into the cells.

58 (a) **True** In the absence of compensation, the pH falls and the $[HCO_3^-]$ rises.
   (b) **True** This is the Henderson-Hasselbalch equation. The curvature of the line ANC reflects the logarithmic relation between pH and $[HCO_3^-]$.
   (c) **False** In hyperventilation the PCO_2 falls. In the absence of any renal compensation, a respiratory alkalosis produces a movement along line ND (i.e. the opposite of a respiratory acidosis).
   (d) **False** A metabolic (i.e. non-respiratory) acidosis produces a movement along line NC, and the locus reached as a result of compensation will lie below and to the right of C because of the fall in PCO_2 (due to hyperventilation) and a fall in $[HCO_3^-]$ (due to hyperventilation and increased excretion of $[HCO_3^-]$ in the urine).
   (e) **True** Haemoglobin is the main buffer in the blood and its concentration is the main factor determining the slope of this line.

*59 At the beginning of an illness a sample of a patient's systemic arterial blood had the following composition: pH 7·2, PCO_2 38 mm Hg (5·1kPa), plasma [HCO_3^-] 14·3 mmol/l. The illness continued and two days later the situation was: pH 7·36, PCO_2 33 mm Hg (4·4kPa), plasma [HCO_3^-] 18 mmol/l.
   (a) The second blood sample was within normal limits.
   (b) A possible diagnosis was respiratory acidosis.
   (c) A possible diagnosis was non-respiratory alkalosis.
   (d) Respiratory compensation must have taken place between the taking of the first and second blood samples.
   (e) Renal compensation must have taken place between the time of the first and second blood samples.

*60 Persistent vomiting leads to a complicated disorder of acid-base balance, because there is a loss of both H^+ and K^+ from the body:
   (a) The loss of H^+ results in a non-respiratory alkalosis.
   (b) The loss of K^+ results in an intracellular alkalosis.
   (c) The plasma HCO_3^- concentration will be lower than normal.
   (d) Respiration will be depressed.
   (e) An acid urine will be produced.

*61 A loss of water in excess of NaCl can occur in certain situations.
   (a) When this happens, glomerular filtration rate is usually increased.
   (b) Diabetes mellitus would give rise to this condition.
   (c) Diabetes insipidus would give rise to this condition.
   (d) Addison's disease would give rise to this condition.
   (e) Severe sweating would give rise to this condition.

*62 In a primary dehydration (also called a hypertonic contraction), there is a loss of water in excess of salt; some features of this condition are:
   (a) The osmolarity of the intracellular fluid is increased.
   (b) The intracellular fluid volume remains unchanged.
   (c) There is a low rate of production of concentrated urine.
   (d) The rate of salt loss in the urine is increased.
   (e) The body temperature often rises.

*63 In a secondary dehydration (also called a hypotonic contraction), there is a loss of salt in excess of water; some features of this condition are:
   (a) A decrease in intracellular fluid volume.
   (b) The patient complains of severe thirst.
   (c) Haemoconcentration occurs (raised haematocrit and plasma protein concentration).
   (d) There is a good flow of dilute urine.
   (e) Marked peripheral vasoconstriction may occur.

59 (a) **False** The pH is within normal limits, but the PCO_2 and HCO_3^- are not.
   (b) **False** PCO_2 is not raised above normal value ($\sim$ 40 mm Hg) (5·3 kPa).
   (c) **False** The pH of the first blood sample is below normal.
   (d) **True** Respiratory compensation would explain the reduced PCO_2 but renal compensation is required to account for the rise of HCO_3^- concentration.
   (e) **True** As shown by the rise in HCO_3^- concentration.

60 (a) **True** Non-respiratory alkalosis is often called metabolic alkalosis.
   (b) **False** The loss of K^+ from the body depletes the intracellular stores, and H^+ replaces the lost K^+; this results in an intracellular *acidosis.*
   (c) **False** It is raised in a non-respiratory alkalosis (but lowered in a respiratory alkalosis – think about that).
   (d) **True** This occurs because of the raised plasma pH.
   (e) **True** This is not the renal response you would expect to an alkalosis; it results from (b), and it is called paradoxical aciduria.

61 (a) **False** Glomerular filtration is not regulated to compensate for disorders of salt and water balance.
   (b) **True** Because diabetes mellitus can result in an osmotic diuresis.
   (c) **True** Lack of ADH results in excretion of a hypotonic urine.
   (d) **False** Loss of salt in excess of water is the primary problem when there is a deficiency of adrenal cortical hormones (aldosterone in particular).
   (e) **True** Sweat contains NaCl, but is less concentrated than the interstitial fluid.

62 (a) **True** Osmolarity of all fluid compartments is increased.
   (b) **False** Volume of all fluid compartments is decreased.
   (c) **True** Due to high rate of secretion of ADH by posterior pituitary.
   (d) **True** Due to reduced rate of secretion of aldosterone by adrenal cortex.
   (e) **True** This culminates in 'dehydration fever' due to failure of circulation to dissipate heat when blood volume is severely reduced.

63 (a) **False** The intracellular fluid volume is increased because of an osmotic movement of water from ECF when its osmolarity decreases.
   (b) **False**
   (c) **True** Severe because any loss of fluid to the exterior is aggravated by loss to ICF.
   (d) **True** Consequence of (a)
   (e) **True** Required for maintenance of blood pressure in face of greatly diminished blood volume.

*64 Ankle oedema accompanies the following diseases. (Note that it is assumed that each is uncomplicated by other diseases.)
(a) Renal failure (nephrotic syndrome).
(b) Kwashiorkor.
(c) Heart failure.
(d) Myxoedema.
(e) Hypertension.

65 A pouch of the stomach may be made such that it is denervated but still has its normal blood supply. Gastric juice may be collected from such a pouch in the conscious dog. The following are statements about secretion from the pouch:
(a) There is a resting secretion.
(b) A normal secretory response to the sight of food is produced.
(c) Secretion starts about ½ hour after a meal is taken by mouth.
(d) Distension of the pouch increases secretion.
(e) If the pouch is moved to a different part of the body and a new blood supply established, secretion occurs in response to a meal.

66 Total gastrectomy would be expected to lead to the following:
(a) Haemodilution after meals.
(b) Vitamin B_{12} malabsorption.
(c) Grossly reduced iron absorption.
(d) Malabsorption of protein.
(e) Impaired fat absorption.

67 The following are statements about pancreatic function:
(a) Stimulation of the vagus produces a rapid secretion of watery juice from the pancreas.
(b) In cross-circulation experiments when acid is introduced into the duodenum of one animal, pancreatic secretion occurs in the other.
(c) Pancreatic secretion contains enzymes which break down polysaccharides.
(d) Pancreatic acini contain trypsin.
(e) The hormone responsible for provoking the enzyme-rich pancreatic secretion also causes contraction of the gall bladder.

68 The following are statements about patients lacking exocrine pancreatic secretion:
(a) Fat digestion is normal provided bile is still produced.
(b) Protein digestion is inefficient and loss of weight tends to occur.
(c) A bleeding tendency may be present.
(d) Excessive water loss takes place via the gut primarily because of inadequate Na^+ absorption in the ileum.
(e) Fasting blood sugar is elevated.

64 (a) **True** Colloid osmotic pressure falls due to renal loss of plasma proteins.
   (b) **True** Dietary deficiency of essential amino acids results in lack of plasma proteins.
   (c) **True** This results mainly from expansion of the ECF due to excessive aldosterone production, but it is aggravated by raised venous pressures in severe forms of congestive cardiac failure.
   (d) **False** A thickening of the subcutaneous tissue is seen in thyroid deficiency as a result of changes in the ground substance.
   (e) **False** Oedema is not a feature of hypertension unless there are complications (e.g. heart failure).

65 (a) **True** There is normally a low level of circulating gastrin.
   (b) **False** This response is neurally mediated via the vagus.
   (c) **True** The delayed response is humorally mediated, and is due to gastrin release.
   (d) **True** The secretory response to distension is largely a local response.
   (e) **True** Humoral influences are still effective becase they act via the blood stream.

66 (a) **False** Haemoconcentration is more likely. After gastrectomy the presence of large volumes of hypertonic digestion products in the duodenum draws water into the gut lumen (dumping syndrome).
   (b) **True** There would be a lack of intrinsic factor.
   (c) **False** It might be reduced slightly due to lack of conversion of Fe^{3+} to Fe^{2+}.
   (d) **True** There is submaximal stimulation of duodenal hormone release (eg secretin and CCK) and hence submaximal stimulation of pancreatic exocrine stimulation.
   (e) **True** See answer to (d).

67 (a) **False** That is typical of the effect of secretin. Vagal stimulation produces a scanty enzyme-rich juice.
   (b) **True** Acid stimulates the release of secretin into the blood stream from the duodenum.
   (c) **True** Amylases.
   (d) **False** Trypsin is stored and secreted in an *inactive* form, trypsinogen. (Activation of trypsin in the gland causes digestion of the pancreas!)
   (e) **True** It was once thought there were two local hormones – cholecystokinin and pancreozymin – but they are now known to be the same thing (CCK–PZ).

68 (a) **False** Pancreatic lipases are required in addition to bile salts to break down the triglycerides.
   (b) **True** Pancreatic proteases and peptidases are essential for protein digestion.
   (c) **True** Due to poor absorption of fats and fat-soluble vitamins including vitamin K, which is required for the production of prothrombin.
   (d) **False** Pancreatic secretion has no effect on sodium absorption.
   (e) **False** That would result from loss of endocrine function because of lack of insulin.

69 The following are statements about bile:
   (a) Bile salts are derived from waste products of haemoglobin breakdown.
   (b) Reabsorption of bile salts from the intestine leads to further secretion of bile.
   (c) A certain concentration of bile salt is required before normal fat absorption
       can take place.
   (d) Bile is concentrated in the gall bladder.
   (e) Active transport of NaCl out of the gall bladder is the mechanism by which the
       bile is concentrated.

*70 The following are statements about jaundice.
   (a) In obstructive jaundice the faeces are pale and the urine is dark.
   (b) Obstructive jaundice is commonly the result of gall stones blocking the cystic
       duct.
   (c) Hepatocellular jaundice leads to an excessive excretion of bilirubin by the
       kidney.
   (d) Haemolytic jaundice does not lead to changes in the faecal content of ster-
       cobilinogen.
   (e) The production of bilirubin from haemoglobin does not take place in liver
       failure.

71 The following are statements about fat absorption:
   (a) If fats are shaken with water at 37°C, the droplet size is greater when bile salt is
       also present.
   (b) Chylomicra are small droplets of fat found in the small intestinal lumen.
   (c) Most bile salt is absorbed in the duodenum.
   (d) The breakdown products of dietary fats are resynthesised in the intestinal cells
       and pass into the lacteals.
   (e) Deficiencies of fat absorption can lead to poor absorption of vitamins of the B
       group.

72 The following are statements about iron metabolism.
   (a) The main excretory route for iron is via cells shed from the intestinal mucosa.
   (b) Destroyed red cells provide the main immediate source of plasma iron.
   (c) Apoferritin production depends on the plasma iron concentration.
   (d) Apoferritin present in the intestinal mucosal cells prevents iron from gaining
       access to the circulation by combining with it.
   (e) Most of the iron in the plasma is in the free form.

69 (a) **False** You are getting confused with bile pigments, which are the products of haemoglobin metabolism.
   (b) **True** This is the 'enterohepatic' circulation of bile.
   (c) **True** At a sufficient concentration of bile (the critical micellar concentration) water soluble aggregates of bile salts, free fatty acids, monoglycerides, cholesterol and fat-soluble vitamins are formed in the gut.
   (d) **True** About 10 to 20 fold.
   (e) **True** Water moves isosmotically with it.

70 (a) **True** Lack of bile makes the faeces pale, but conjugated bile is excreted in the urine.
   (b) **False** The cystic duct connects the gall bladder to the common bile duct; blockage of the cystic duct does not therefore cause jaundice.
   (c) **False** Damage to the liver leads to failure of conjugation and no renal excretion of bilirubin can occur.
   (d) **False** Stercobilinogen is increased because excessive haemolysis leads to excessive bile pigment excretion in the faeces.
   (e) **False** It is *conjugation* that is impaired in liver failure.

71 (a) **False** The bile salts act as emulsifying agents so droplet size is smaller.
   (b) **False** They are found in mesenteric lymphatics and in the blood stream after a fatty meal.
   (c) **False** In the terminal ileum by active transport.
   (d) **True**
   (e) **False** The fat soluble vitamins are A, D, E and K. If you cannot remember them, invent a mnemonic.

72 (a) **True**
   (b) **True**
   (c) **True**
   (d) **True** Increased plasma iron leads to the production of more apoferritin, which combines with iron and is subsequently shed in the lumen.
   (e) **False** Most of the iron is combined with a carrier protein transferrin.

73 The graphs show the transport rate of individual sugars across a membrane. Transport is occurring from a solution containing the sugar both alone and in the presence of a fixed concentration of other sugars into a solution containing no sugar. Which of the following statements are consistent with the data?

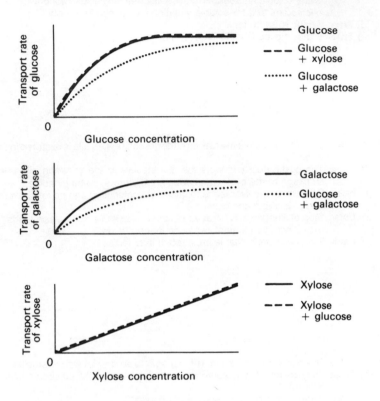

(a) Glucose is transported actively.
(b) Galactose is transported actively.
(c) Xylose is transported actively.
(d) Glucose and galactose compete for the same carrier mechanism.
(e) Glucose and xylose compete for the same carrier mechanism.

*74 The following are statements about diseases affecting the intestine.
   (a) In gluten enteropathy the villi of the jejunum are flattened and reduced in number.
   (b) Gastrin-secreting tumours are sometimes found in the pancreas.
   (c) A failure of both pancreatic and biliary secretion is needed before steatorrhoea develops.
   (d) Cholera toxin leads to excessive secretion of NaCl and water into the intestinal lumen.
   (e) Megaloblastic anaemia is commonly associated with diseases that affect the terminal ileum (e.g. Crohn's disease).

73 (a) **True** A limiting rate is observed.
   (b) **True** For same reason.
   (c) **False** Rate of transport increases in proportion to lumen concentration.
   (d) **True** Presence of one reduces the rate of transport of the other.
   (e) **False** Presence of xylose makes no difference to rate of transport of glucose.

74 (a) **True** Absorption is impaired.
   (b) **True** This leads to excessive production of HCl (Zollinger-Ellison syndrome).
   (c) **False** A failure of either leads to malabsorption of fats.
   (d) **True** Normal NaCl absorption is reversed by a serosal-to-mucosal Cl^- pump, activated by cAMP. Glucose-dependent Na reabsorption is unaffected.
   (e) **True** Vitamin B_{12} is absorbed in the terminal ileum.

75 The following are statements about the antidiuretic hormone (ADH):
   (a) ADH is transported from hypothalamic nuclei to the neurohypophysis via the hypophyseal portal system.
   (b) ADH increases the permeability of the collecting tubules of the kidney to water.
   (c) A high rate of secretion of ADH can lead to complete cessation of urine production.
   (d) A fall in the osmolarity of the blood supplying the hypothalamus is a powerful stimulus for ADH secretion.
   (e) A fall in blood volume results in an increase in ADH secretion.

76 Aldosterone production is likely to be increased:
   (a) If the osmolarity of the extracellular fluid rises.
   (b) After a severe haemorrhage.
   (c) If plasma [K] rises.
   (d) In hypopituitarism.
   (e) Following administration of a carbonic anhydrase inhibitor.

*77 Adrenal cortical insufficiency (Addison's disease) is often accompanied by a low plasma Na concentration and raised plasma K concentration, low blood glucose, pigmentation of the skin, and a low systemic arterial blood pressure:
   (a) The low plasma Na concentration is due largely to failure of the distal convoluted tubules of the kidney to reabsorb sodium.
   (b) The raised plasma K concentration is due to excessive haemolysis of red blood cells.
   (c) The low blood glucose is due to lack of glucocorticoid secretion.
   (d) The pigmentation of the skin is due to the decreased rate of production of ACTH.
   (e) The low blood pressure is due to the increase in extracellular fluid volume.

75 (a) **False** It is transported via neurosecretory granules along the hypothalamo-neurohypophyseal tract. The portal system goes to the anterior pituitary (adenohypophysis)
   (b) **True**
   (c) **False** The upper limit of urine concentration is that of the interstitial concentration at the tip of the renal papillae and is about 1200 mosmol/l. The normal solute load at maximum urine concentration requires about 600 ml urine in 24 hr.
   (d) **False** A *rise* is required.
   (e) **True** Blood volume is monitored by receptors in the walls of the great veins and the atria. A fall in blood volume causes a neurally-mediated release of ADH, in addition to the release of aldosterone. The latter then increases plasma Na concentration which thus raises plasma osmolarity and augments the release of ADH.

76 (a) **False** A *decrease* in [Na] is one of the stimuli for increased aldosterone production.
   (b) **True** The reduction in extracellular volume is the stimulus in this situation.
   (c) **True** The rate of production of aldosterone is directly related to the plasma [K].
   (d) **False** Aldosterone production continues in the absence of anterior pituitary hormones, but is somewhat reduced.
   (e) **True** The action of carbonic anhydrase in the convoluted tubules is necessary for re-uptake of bicarbonate, which is accompanied by Na, from the tubular fluid.

77 (a) **True** Aldosterone affects sodium reabsorption in both proximal and distal tubules, but it is the impairment of absorption in the latter that results in the low plasma Na concentration.
   (b) **False** It is a result of the failure of the aldosterone-dependent secretion of K in exchange for Na absorbed in the distal convoluted tubule.
   (c) **True** Glucocorticoids promote glycogenolysis and gluconeogenesis.
   (d) **False** It is due to an increased rate of production of ACTH or of some trophic hormone closely related to it.
   (e) **False** The deficiency of aldosterone leads to a reduction of ECF volume and the fall in blood volume is partly responsible for the low blood pressure. The lack of glucocorticoids is also important because the arteriolar smooth muscle becomes less responsive to pressor influences.

*78 The following are common features of pan-hypopituitarism:
   (a) Amenorrhoea.
   (b) Increased urine output.
   (c) Excessive growth of body hair.
   (d) Decreased tolerance to cold weather.
   (e) A form of diabetes mellitus.

*79 Would you expect poor calcification of bone to occur in the following situations?
   (a) Cushing's syndrome.
   (b) Hypoparathyroidism.
   (c) Poor dietary intake of cholecalciferol (vitamin D) in black children living in northern areas of Europe.
   (d) Tumours of the thyroid C cells.
   (e) Renal disease.

80 The following are statements about the female reproductive system:
   (a) Both oestrogen and progesterone are necessary for ovulation to take place.
   (b) Oestrogen tends to inhibit the production of FSH by the anterior pituitary gland.
   (c) Fertilization of the ovum by the spermatozoon normally takes place in the uterus.
   (d) Progesterone production is largely under the control of LH.
   (e) Throughout the part of the menstrual cycle that follows ovulation, there is a slight rise in body temperature.

*81 Which of the following would suggest a diagnosis of thyrotoxicosis?
   (a) Pulse rate 65/min.
   (b) Loss of weight.
   (c) Raised plasma triiodothyronine concentration.
   (d) Raised plasma TSH concentration.
   (e) Poor appetite.

78 (a) **True** Pan-hypopituitarism means loss of function of both anterior and post-erior pituitary. Gonadotrophins are required for menstruation.
   (b) **True** Due to the absence of ADH.
   (c) **False** Hirsuitism is characteristic of excess corticoid secretion e.g. Cushing's syndrome.
   (d) **True** The thyroid response would be abolished because of TSH lack and the general response to stress is diminished because of ACTH lack.
   (e) **False** *Excessive* secretion of growth hormone can produce this.

79 (a) **True** High steroid output leads to poor collagen synthesis. Since collagen is being broken down continuously, albeit slowly, this leads to dissolution of the bone matrix.
   (b) **False** Poor calcification occurs in hyperparathyroidism.
   (c) **True** Vitamin D is required for normal calcium absorption. When dietary intake is inadequate, synthesis in the skin by UV irradiation of calciferol from the sun becomes important. Pigmentation reduces UV penetration and in northern areas there is little sun anyway.
   (d) **False** These produce calcitonin. Calcitonin enhances bone calcification.
   (e) **True** The active form of vitamin D is produced in the kidney. Renal disease can lead to 'renal rickets' in children and osteomalacia in adults.

80 (a) **False** FSH and LH are needed.
   (b) **True**
   (c) **False** It normally takes place in the uterine tubes.
   (d) **True** LH acts on the corpus luteum.
   (e) **True** Occasionally there is a transient fall at the time of ovulation before the sustained rise.

81 (a) **False** Basal metabolic rate is raised in thyrotoxicosis and one sign of this is a high heart rate. The rate quoted is within the normal range.
   (b) **True** The body's food reserves become depleted if the rise in metabolic rate is not matched by an adequate increase in food intake.
   (c) **True** Triiodothyronine and thyroxine (tetraiodothyronine) are the thyroid hormones, and generally both are produced in excess in thyrotoxicosis.
   (d) **False** When the thyroid is producing excessive amounts of hormones, the level of TSH is depressed by a negative-feedback action.
   (e) **False** Increased appetite (see b).

*82 The following would be consistent with a diagnosis of diabetes mellitus:
    (a) A glucose tolerance test as in the diagram.

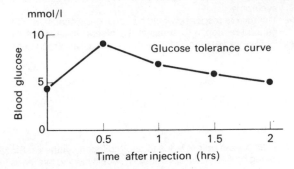

    (b) The following acid-base picture in the blood: pH 7·15; PCO_2, 35 mm Hg (4·6 kPa); HCO_3^-, 15 mmol/l.
    (c) Ketoaciduria.
    (d) Loss of weight.
    (e) Nocturia.

83 Net movement of a substance by simple diffusion:
    (a) *Can* occur against a concentration gradient.
    (b) Can occur against an electrochemical gradient.
    (c) Can be slowed by a factor of at least two by lowering the temperature 10°C.
    (d) Is unaffected by metabolic poisons.
    (e) Shows saturation properties.

84 Net movement of a substance by active transport:
    (a) Can occur against a concentration gradient.
    (b) Can occur against an electrochemical gradient.
    (c) Can be slowed by a factor of at least two by lowering the temperature 10°C.
    (d) Is unaffected by metabolic poisons.
    (e) Shows saturation properties.

85 The following are examples of active transport:
    (a) Chloride shift between red blood cells and the plasma.
    (b) Sodium reabsorption in the distal tubules of the kidney.
    (c) Movement of oxygen from pulmonary alveoli into the blood.
    (d) Uptake of calcium by the sarcoplasmic reticulum of muscle.
    (e) Oxygen movement within a muscle fibre.

86 Ultrafiltration is responsible for the fluid movement that takes place in the following processes:
    (a) Concentration of bile.
    (b) Salivation.
    (c) Glomerular filtration.
    (d) Sweating.
    (e) Interstitial fluid production.

82 (a) **False** The blood glucose was initially within normal fasting limits, and it returned to within the normal range soon after the injection.
　　(b) **True** The low plasma pH and low bicarbonate are consistent with the production of ketoacids in diabetes mellitus. The PCO_2 is low because there has probably been some respiratory compensation for the non-respiratory acidosis.
　　(c) **True** Ketoacids formed by catabolism of fats are excreted in the urine.
　　(d) **True** The lack of insulin reduces glucose entry into the cells. Gluconeogenesis therefore occurs and the consumption of fats and proteins leads to a loss in body weight.
　　(e) **True** Blood sugar is raised above the renal threshold, and the excretion of glucose in the urine results in an osmotic diuresis. This can be sufficiently severe to necessitate emptying the bladder during the night (nocturia).

83 (a) **True** But only if it is changed and there is an electrical gradient in its favour.
　　(b) **False**
　　(c) **False** It is slowed down much less than this (about 2·5% per °C).
　　(d) **True**
　　(e) **False**

84 (a) **True** ⎫
　　(b) **True** ｜ These are the properties required
　　(c) **True** ⎬ if a transport system is to be considered
　　(d) **False** ｜ active.
　　(e) **True** ⎭

85 (a) **False** The electrical gradient due to HCO_3^- loss from red cells favours movement of Cl^- into the cells.
　　(b) **True**
　　(c) **False**
　　(d) **True** The uptake of calcium from sarcoplasm is against a concentration gradient.
　　(e) **False** This is an example of facilitated diffusion, in which myoglobin is the carrier.

86 (a) **False** This is by active transport of Na and Cl out of the gall bladder as an isosmotic solution.
　　(b) **False** This is by active ion transport into the acinar lumen (probably Na) followed by isosmotic water transfer.
　　(c) **True**
　　(d) **False** Sweating is a similar process to salivation.
　　(e) **True**

N.B. Qs 87 and 88 are searching questions which have been included to test your understanding rather than as examples of examination questions.

87 Suppose we have two compartments separated by a membrane (properties to be specified below). Compartment 1 is filled with a weak solution of NaCl and compartment 2 is filled with a solution of KCl at the same concentration.

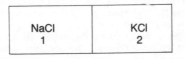

NaCl 1	KCl 2

   (a) If the membrane is equally permeable to Na, K and Cl, redistribution of ions will occur until the two compartments have the same composition.
   (b) If the membrane is more permeable to K and Cl than to Na, the Na concentration in compartment 1 will be greater than that in compartment 2 when all net movement of ions ceases.
   (c) If the two compartments have rigid walls and the membrane is permeable to K and Cl but not to Na, compartment 1 will be positive with respect to compartment 2 when all net movement of ions ceases.
   (d) If the two compartments have freely distensible walls and the membrane is permeable to K and Cl but not to Na, compartment 1 will expand and compartment 2 will shrink.

88 Suppose we have two compartments with rigid walls separated by a membrane which is permeable to K and Cl but not to Na. Compartments 1 and 2 are filled with solutions of NaCl and KCl, respectively, each at a concentration of 12 mmol/l. When all net movement of ions has ceased:
   (a) The concentration of KCl in compartment 2 will have fallen to about 8 mmol/l.
   (b) There will be a potential difference of about 17 mV between the two compartments.
   (c) The osmolarity of compartment 1 will be about twice that of compartment 2.
   (d) The membrane between the two compartments will be subjected to a transmural pressure of about 270 mm Hg (36 kPa).

89 The resting potential across the nerve membrane:
   (a) Depends on the ratio of the potassium concentrations inside and outside the cell.
   (b) Is positive inside with respect to outside.
   (c) Is of the order of 0.1 volt.
   (d) Decreases in magnitude during prolonged anoxia.
   (e) Is greater the larger the diameter of the fibre.

87 (a) **True**

   (b) **False** If the membrane is even slightly permeable to all three ions the compositions of the solutions in the two compartments will eventually be the same.

   (c) **True** K will diffuse down its concentration gradient from 2 to 1, setting up a voltage gradient across the membrane which causes Cl to follow. Eventually an equilibrium is reached when $[K]_1 \times [Cl]_1 = [K]_2 \times [Cl]_2$. The tendency for K and Cl to move down their concentration gradients is then opposed by a voltage difference between the two compartments.

   (d) **True** The ion movements described in (c) will occur, but this results in osmotic imbalance between the two compartments. If the walls are distensible, water will therefore move from compartment 2 to compartment 1.

88 (a) **True** The events described in 87(c) will take place and a Donnan equilibrium will be established – i.e.

$$[K]_1 \times [Cl]_1 = [K]_2 \times [Cl]_2$$

If the fall in concentration in compartment 2 is denoted by n mmol/l for K and Cl, then substituting in the above equation gives:

$$n \times (12 + n) = (12 - n) \times (12 - n)$$

From this $n = 4$ and the final concentration of KCl in compartment 2 will therefore be 8 mmol/l. (Note that this approach involves some approximations, but these have a negligible effect on the final answer.)

   (b) **True** From the calculation in (a) it can be seen that the concentration ratio for $[K]_2/[K]_1$ and $[Cl]_1/[Cl]_2$ is 2. For monovalent ions at 25°C the Nernst equation gives an equilibrium potential (mV) of

$$58 \times \log_{10} \text{(concentration ratio)} = 17.5 \text{ mV}.$$

Compartment 1 will be positive with respect to 2.

   (c) **True**

   (d) **True** There will be a difference in osmolarity of about 16 mosmol/l between the two compartments. 1 osmole in 22·4 l exerts a pressure of 760 mm Hg (101 kPa). Hence there will be a transmural osmotic pressure difference of about 270 mm Hg (36 kPa).

89 (a) **True** It is proportional to $\log \dfrac{[K]_{in}}{[K]_{out}}$ for small changes in $[K]_{out}$ (see 88b).

   (b) **False** Negative inside.

   (c) **True**

   (d) **True** The sodium pump, which maintains the gradient of K and Na across the cell membrane depends on metabolic energy.

   (e) **False** *Conduction velocity* of action potential depends on diameter in this way.

90 This is a schematic diagram of an action potential recorded from an axon with an intracellular electrode;

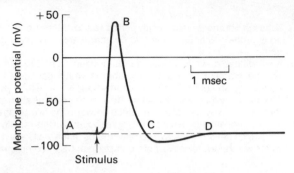

(a) The resting membrane potential (A) is close to the potassium equilibrium potential.
(b) The rapid change in membrane potential from A to B is due to a self-regenerative rise in the permeability of the cell membrane to potassium ions.
(c) The membrane potential reaches a peak (B) that is close to the sodium equilibrium potential.
(d) The repolarisation of the cell membrane from B to C is due to the extrusion of sodium ions from the axon by the sodium pump.
(e) The absolute refractory period ends at D, when the membrane potential finally reaches its previous steady level.

91 The left-hand diagram shows a frog sciatic nerve lying across a number of electrodes. A and B are used for stimulating and C and D for recording. The right hand diagram shows a typical recorded action potential.

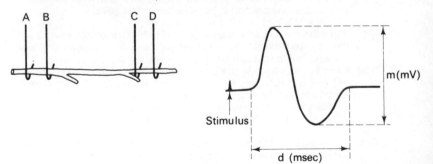

(a) The first deflection on the recording occurs when electrode C is negative with respect to D.
(b) The magnitude of the recorded action potential (m) will be independent of the distance between electrodes C and D.
(c) The duration of the recorded action potential (d) will be independent of the distance between electrodes C and D.
(d) The duration of the recorded action potential will depend on the distance between B and C.
(e) The recorded action potential can be made monophasic by crushing the nerve (or applying local anaesthetic) at A.

90 (a) **True**
   (b) **False** To Na ions.
   (c) **True**
   (d) **False** It is due to a turning off of the increased permeability to Na and an increase in K permeability, thus restoring the membrane potential close to the K equilibrium potential.
   (e) **False** It ends just before C.

91 (a) **True** The first deflection occurs when the wave of depolarisation, travelling from B along the nerve, reaches electrode C.
   (b) **False** If the electrodes are close together, depolarization may reach D at a time when there is still depolarization under C. There would then be a smaller potential difference between the electrodes than if D were at a more distant site.
   (c) **False** You should be able to work this out from the explanation given in (b).
   (d) **True** The further apart the stimulating and recording electrodes, the more dispersed are the action potentials in fast and slow conducting fibres when they reach C, and hence the longer the wave duration.
   (e) **False** By crushing or local anaesthetic at D.

92 The following are statements about cell-to-cell transmission:
  (a) An impulse in an axon that releases excitatory transmitter invariably produces an action potential in the postsynaptic neurone.
  (b) The amount of transmitter released at the neuromuscular junction depends on the calcium ion concentration in the surrounding medium.
  (c) Inhibitory transmitter produces an increase in conductance of the post-synaptic membrane to K^+ and Cl^-.
  (d) The generation of an action potential in a skeletal muscle fibre by neuro-muscular transmission requires the arrival of several nerve impulses in rapid succession at the motor end plate.
  (e) Transmission across a synapse can occur in both directions.

93 The following are statements about the autonomic nervous system:
  (a) The ratio of the number of preganglionic : postganglionic fibres is about 20:1.
  (b) The adrenal medulla secretes hormones with actions like those of the post-ganglionic nerves of the sympathetic nervous system.
  (c) The highest centre involved in integration of the autonomic nervous system is in the medulla oblongata.
  (d) Transmission velocity in postganglionic autonomic nerves is about the same as that in somatic motor nerves.
  (e) The effects of activity in parasympathetic nerves are more localised than those of sympathetic activity.

94 Acetylcholine is the chemical transmitter at:
  (a) All neuromuscular junctions in the somatic nervous system.
  (b) All post-ganglionic sympathetic endings.
  (c) All autonomic ganglia.
  (d) All post-ganglionic parasympathetic effector endings.
  (e) Any sites that are blocked by atropine.

95 The following processes are brought about by activation of the parasympathetic nerve fibres:
  (a) Defaecation.
  (b) Micturition.
  (c) Sweating.
  (d) Ejaculation of semen.
  (e) Dilation of the pupil.

96 The following are statements about vasomotor nerves:
  (a) Sympathetic vasoconstrictor nerves supply the smooth muscle in the walls of arterioles almost everywhere in the body.
  (b) The transmitter released by vasoconstrictor nerves is noradrenaline.
  (c) There are also sympathetic vasodilator nerves that supply arterioles in striated muscles.
  (d) The transmitter released by these vasodilator nerves is adrenaline.
  (e) Both types of vasomotor nerves are involved in the baroceptor reflex.

92 (a) **False** While this is true for skeletal neuro-muscular transmission, it is rarely true at synapses in the nervous system.
   (b) **True** More is released when the $[Ca^{++}]$ is high.
   (c) **True** This would tend to shunt excitatory effects and may hyperpolarise the membrane.
   (d) **False** One impulse is sufficient for extrafusal muscle fibres.
   (e) **False** Unidirectional transmission is imposed by anatomical organisation.

93 (a) **False** There are always more postganglionic than preganglionic fibres. The ratio varies from 1 preganglionic to 1 ot 2 postganglionic fibres in some parasympathetic ganglia to 1: 20 or 30 in some sympathetic ganglia.
   (b) **True** Adrenaline and noradrenaline.
   (c) **False** The hypothalamus is the most important higher integrating centre.
   (d) **False** Much less. Postganglionic autonomic fibres are small (0·5 to 1 μm diameter) and unmyelinated.
   (e) **True** Work out why from the anatomy and mode of destruction of transmitters.

94 (a) **True**
   (b) **False** Noradrenaline is the transmitter at most of these endings.
   (c) **True**
   (d) **True**
   (e) **True** Atropine blocks the effects of acetylcholine released at postganglionic parasympathetic junctions (muscarinic effects of acetylcholine).

95 (a) **True**
   (b) **True**
   (c) **False** Sympathetic cholinergic fibres.
   (d) **False** Sympathetic fibres.
   (e) **False** Sympathetic fibres.

96 (a) **True**
   (b) **True**
   (c) **True**
   (d) **False** The transmitter is acetylcholine but adrenaline released from the adrenal medulla does act on β receptors to cause vasodilatation in the muscle vessels.
   (e) **False** Changes in sympathetic vasoconstrictor tone are alone responsible. Sympathetic cholinergic vasodilator nerves are not involved.

97 The following are statements about the role of calcium in nerve and muscle:
   (a) Calcium provides a link between excitation and contraction in all types of muscle.
   (b) The Ca^{++} concentration is much lower in the cytoplasm of nerve and muscle fibres than it is in the extracellular fluid.
   (c) A reduction in the extracellular Ca^{++} concentration decreases the excitability of nerve and muscle fibres.
   (d) An increase in the extracellular Ca^{++} concentration results in an increase in the strength of contraction of skeletal muscle fibres.
   (e) More transmitter is released from nerve terminals by an action potential when the extracellular Ca^{++} concentration is increased.

98 The following are statements about extrafusal motor units:
   (a) From the point of view of neural control, a motor unit is the functional unit in a muscle.
   (b) There is a considerable overlap between one motor unit and another in the muscle fibres they control: i.e. a given muscle fibre is innervated by branches from several motor axons.
   (c) The only way available to the nervous system for increasing the strength of a muscle contraction is to recruit more motor units.
   (d) Smooth voluntary movements result from the fact that active motor units always produce fused tetanic contractions.
   (e) An action potential in a nerve axon will normally excite every muscle fibre making up that motor unit.

99 Myogenic tone is:
   (a) A continuous discharge of impulses in vasomotor fibres of the sympathetic nervous system.
   (b) A state of sustained contraction in postural skeletal muscles.
   (c) A continuous discharge from annulo-spiral endings of muscle spindles.
   (d) A sustained contraction that occurs in smooth muscles.
   (e) A maintained contraction that occurs in any type of muscle in the absence of electrical activity in the muscle cells.

100 The following are statements about the properties of cardiac muscle:
   (a) Action potentials propagate from one muscle fibre to another.
   (b) The action potential lasts almost as long as the mechanical response.
   (c) Normal heart beats are neurogenic in origin.
   (d) A fused tetanic response can be produced by repetitive stimulation.
   (e) Spontaneous contractions occur in the muscle due to the presence of pacemaker cells.

97 (a) **True**
   (b) **True** It is actively taken up by sarcoplasmic reticulum in muscle, and extruded from nerve.
   (c) **False** Ca^{++} stabilizes the membrane so a fall in extracellular Ca^{++} increases excitability. At concentrations below normal, but higher than in overt tetany, the facial nerve can be activated merely by tapping the overlying skin (a sign of latent tetany).
   (d) **False** Ca^{++} required for excitation-contraction coupling in skeletal muscle is provided by internal stores and Ca^{++} is recycled within the cell.
   (e) **True**

98 (a) **True**
   (b) **False** In most mammalian muscles a given muscle fibre is innervated only from a single motor axon.
   (c) **False** It is one way; the other is to increase the frequency of nerve firing which increases the force of contraction by producing a tetanus.
   (d) **False** Firing rates are insufficient to produce a fused tetanus in extrafusal motor units. The smooth contractions are due to the compliance of tendons, etc., in the muscle and the fact that different motor units contract asynchronously.
   (e) **True**

99 (a) **False** This is vasomotor tone.
   (b) **False** This is dependent on neural activity.
   (c) **False**
   (d) **True** This results from spontaneous electrical activity in the smooth muscle cells.
   (e) **False** The appropriate name for this is a contracture.

100 (a) **True**
   (b) **True**
   (c) **False** They are myogenic, arising in pacemaker tissue. The autonomic nerve supply can affect the *rate* of firing in pacemaker cells.
   (d) **False** The long action potential means that the cell membrane is refractory throughout most of the mechanical response.
   (e) **True**

64

First published 1978
by Edward Arnold (Publishers) Ltd
41 Bedford Square, London WC1B 3DQ

Reprinted 1979, 1982, 1985

British Library Cataloguing in Publication Data
Bindman, Lynn
    Multiple choice questions in physiology.
    1. Physiology – Miscellanea
    I. Title II. Jewell, Brian III. Smaje,
    Laurence
    612'.0076   QP40

ISBN 0–7131–4314–2

Printed in Great Britain by
Pitman Press, Bath